Contents

1 Farm Life

Farming is hard work. It is made up of many jobs. Nowadays many jobs are done by machines. In 1700 most jobs were done by men and women, horses and oxen.

Source A

'The farmer is thought of chiefly as a supplier of food. However, for hundreds of years other industries have relied on the farmer - not only industries like milling and brewing. All textile industries needed starch. But the largest industry in Britain until about 1800 spun and wove wool. Tallow (made of grease from sheep's wool) formed the basis of most candles. Leather made footwear, horse harness, bags and parts of some machines.'

From *Britain Yesterday and Today* by W. M. Stern, 1962.

Source B

'In November the good farmer watches over the thresher otherwise he will steal something. In December he sends his bailiff to collect all yokes, forks and farm tools so he can put them in good repair. In January there is hedging and ditching. In February and March he must see well to his sheep. In April it is time to get out the hop poles and sell bark to the tanner before the trees are felled. Elm and ash are put by for ploughs and carts, hazel for forks, sallow for rakes and thorn for threshing flails. In May there is weeding and children must be hired to pick up stones. In June there is washing and shearing sheep. Then in July comes the hay harvest when everyone works all hours. In August comes the corn harvest and in September the fruit to be gathered, wool sold, logs stacked for the winter; salt fish bought in town for Lent next year. Then there is ploughing and later sowing. . . .'

From *Elizabethan Life in Town and Country* by M. St. Clare Byrne, 1925.

Source C

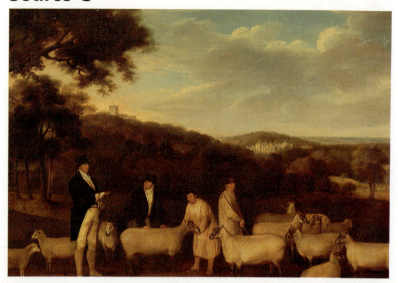

Thomas Coke, the first Earl of Leicester, lived in Norfolk. His house (shown in the picture) was called Holkham Hall. He owned many farms. He encouraged his good farmers by giving them long leases. (This meant the farmers knew they had the farms for many years, so they would spend their own time and money improving them.) Coke encouraged new crops to be grown. He also held sheep shearings every year. They were like agricultural shows. Farmers met to talk and exchange new farming ideas.

Source D

'The harvest month was up on Saturday night; but the barley, some oats, beans and spring tares (crops) are still to be cut, and come in.

In the afternoon I intended to have carried some of the barley from stooks (bundles), but just as we were about to begin, the rain set in. This is the third or fourth time it has happened. Had the fine weather continued but six hours longer, we should have done it. It is lucky we had it in stooks however, for that preserved its sap and colour.'

From *The Minutes of Agriculture 1778* by William Marshall.

Agriculture Since 1700

Fiona Reynoldson

HEINEMANN EDUCATIONAL

Heinemann Educational
a division of Heinemann Educational Books Ltd,
Halley Court, Jordan Hill, Oxford OX2 8EJ

OXFORD LONDON EDINBURGH
MELBOURNE SYDNEY AUCKLAND
IBADAN NAIROBI GABORONE HARARE
KINGSTON PORTSMOUTH NH (USA)
SINGAPORE MADRID BOLOGNA ATHENS

© Fiona Reynoldson 1990

First published 1990

British Library Cataloguing in Publication Data

Reynoldson, Fiona
  Agriculture since 1700.
  1. Agriculture, history
  I. Title
  630.9i

ISBN 0-435-31092-5

Designed and produced by
VAP Publishing Services, Oxford

Printed in Spain by Mateu Cromo

Acknowledgements

The author and publisher would like to thank the following
for permission to reproduce photographs in the Units
indicated:

Aerofilms Ltd: 4A
Barnaby's Picture Library: 22D
Birmingham City Council Museums and Art Gallery: 13G
Bridgeman Art Library/Private Collection: 9G
Cider Museum, Hereford: 13C
Coke Estates Ltd, courtesy of Viscount Coke and the
  Trustees of the Holkham Estate: 1C
Dorset County Museum: 14D
The Fine Art Society: 11H
Derek Forss: 7A, 21A
Imperial War Museum: 15C, 16A and B, 17D, 19B and G
Institute of Agricultural History and Museum of English
  Rural Life, University of Reading: 3B and F, 4B, 5A and
  D, 6E, 7G, 8C, 9B, 10C, 12A, C and E, 14A, 15E, 16E,
  17E, 18A, B and E, 20B, 21D, 22E, and cover
Institute of Agricultural History and Museum of English
  Rural Life, University of Reading/*Farmers Weekly*: 2G
Collection of Linda Mackie/Out of the West Publishing: 15F
The Mansell Collection: 2B, 8D, 10A
Pitt Rivers Museum: 6C
Punch Publications: 13A, 14E
Anne Ronan Picture Library: 10D
Rothamsted Experimental Station: 10F
Stoke-on-Trent Museum: 9A
Suffolk Record Office: 7B
T. A. Wilkie Photo Library: 22G

We have been unable to contact the copyright holder of
the painting which appears on page 3 and would be
grateful for any information that would enable us to do so.

The author and publishers would also like to thank David &
Charles Publishers for permission to reproduce the maps
and diagrams which appear on pages 4, 6 and 21, taken
from *A History of Farm Buildings in England and Wales* by
Nigel Harvey, 1984 and Grafton Books for permission to
reproduce the illustration on page 11, from *History Alive 2*
by Peter Moss.

Details of Written Sources

In some sources the wording or sentence structure has
been simplified to make sure that the source is accessible.

Joseph Arch, *Ploughtail to Parliament*, Cresset Library,
  1986: 6F
Harold Bennett, *Farming with Steam*, Shire Publications,
  1974: 12F
Ronald Blythe, *Akenfield: Portrait of an English Village*,
  Allen Lane, 1969: 11C
Asa Briggs, *The Age of Improvement*, Longman, 1960: 10B
A. J. Cairncross, *Home and Foreign Investment, 1870–1913*,
  Cambridge University Press, 1953: 13G
J. D. Chambers and G. E. Mingay, *The Agricultural
  Revolution*, Batsford, 1966: 2D, 8A
Rev. Andrew Clark, *Echoes of the Great War, 1914–19*,
  Oxford University Press, 1985: 17A
J. S. Collis, *The Worm Forgives the Plough*, Barrie &
  Jenkins, 1946: 12D, 19E
R. J. Cootes, *Britain Since 1700*, Longman, 1982: 6E
J. G. Crowther, *The Story of Agriculture*, Hamish Hamilton,
  1958: 5B
Nigel Harvey, *A History of Farm Buildings in England and
  Wales*, David & Charles, 1984: 2E
W. G. Hoskins, *One Man's England*, BBC, 1978: 22G
W. H. Hudson, *A Shepherd's Life*, Methuen, 1910: 8E
David Jones, *Rural Crime and Protest*, Routledge & Kegan
  Paul, 1981: 9D, 9F
Charles Kightly, *Country Voices*, Thames & Hudson,
  1984: 16D
Peter Mathias, *The First Industrial Nation*, Methuen,
  1969: 13D
Trevor May, *The Economy 1815–1914*, Collins, 1972:
  12B, 15B, 15D
Gordon E. Mingay (Ed.), *The Victorian Countryside*,
  Routledge & Kegan Paul, 1981: 11B, 11E
P. Pagnamenta and R. Overy, *All Our Working Lives*, BBC,
  1984: 16C, 17B, 17D, 18C, 19A, 20A, 21A, 22B
Roy Palmer, *The Painful Plough*, Cambridge University
  Press, 1972: 14C
Lawrence Rawstorne, *Gamonia or the Art of Preserving*,
  Ackermann, 1837: 9E
M. St. Clare Byrne, *Elizabethan Life in Town and Country*,
  Methuen, 1925: 1D
W. M. Stern, *Britain Yesterday and Today*, Longman,
  1962: 1A
Flora Thompson, *Lark Rise to Candleford*, Oxford
  University Press, 1947: 3C, 11A
Shirley Toulson, *The Drovers*, Shire Publications, 1980:
  7C, 7F
Ben Wicks, *No Time to Wave Goodbye*, Bloomsbury,
  1988: 19C
R. M. Williams, *British Population*, Heinemann Educational
  Books, 1972: 2C
L. Woodward, *The Age of Reform*, Oxford University
  Press, 1962: 10E
'The Killing Fields', *Radio Times*, BBC Publications,
  1989: 22A

Source E

The Farmyard, painted about 1790.

Questions

1 Describe the different types of jobs shown in Source E.

2 Make a list of all the farm jobs shown or written about in the other sources.

3 What time of year is it in Source E? How can you tell?

4 What sorts of food are being produced in Source E?

5 What is being preserved in Source E for winter food?

6 What evidence is there in Source E to support the statements in Source A?

7 What evidence is there in the sources that the weather is very important to farmers?

2 The Old Farming

Over one thousand years ago the Saxons settled in Britain. They were farmers. They built houses close together. This made a village. They cut down trees, dug up the tree stumps and burnt bushes. In this way they made big, open fields around the village. The farmers divided the fields into strips. Each farmer owned some strips in each field.

Each year one field was planted with wheat (for bread) another with barley (for beer). The third was not planted with anything. It was left fallow to rest. Weeds and grass grew on the fallow field. Animals grazed it, and manured it. This sort of farming went on for hundreds of years in many parts of Britain.

By 1750 the population of Britain was growing fast. More food was needed to feed more people.

Source A

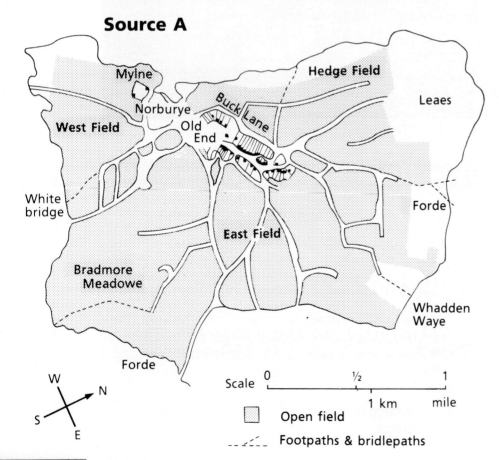

The village of Padbury in Buckinghamshire, from a map of 1591.

Source B

Every autumn many animals were slaughtered because there was not enough food to keep them through the winter.

Source C

'The population of England and Wales grew from about 5.8 million in 1701 to 11.9 million in 1801. Growth rates went up and down, but those later in the century, after 1741, speeded up and increased.'

From *British Population* by R. M. Williams, 1972.

Source D

'Perhaps the most striking weakness of the old farming in open fields was the annual fallowing of about a third of the open field land.'

From *The Agricultural Revolution 1750–1850* by J. D. Chambers and G. E. Mingay, 1966.

Source E

> By 1800 London alone offered a market of nearly a million people.
>
> The most obvious way of increasing the supply of food was now failing. The reclamation of farmland from forests, marshes and hills was almost finished. By 1820 there was little wild land left worth farming.
>
> The alternative to more farming was better farming.

From *A History of Farm Buildings in England and Wales* by Nigel Harvey, 1984.

Source F

Map showing the main crops grown in different parts of England and Wales in the seventeenth century. Remember all parts of Britain were largely self-sufficient, so all parts grew a mixture of crops and kept a mixture of animals.

Source G

Laxton in Nottinghamshire, the last open-field village in England. Notice that all the farm buildings are in the village.

Questions

1 How many fields are shown in Source A?

2 a Trace the outline of Source F. Then shade in the areas that grew wheat and barley. Those areas were the places where farming was mostly in open fields before about 1750. Some sheep will be included in the shading.
  b What other crops were grown?

3 Read Source C and Source D.
  a Why did people say the old farming was wasteful by 1750?
  b Why didn't the wastefulness matter before 1750?

4 How had farmers increased the supply of food before about 1800? (Look at Source E.)

5 How do you think the meat was preserved in Source B?

6 According to Source E why is more food needed? How does the writer think this food will have to be produced?

3 Enclosure

Some people owned many strips in the open fields. The strips were scattered all over the big, open fields. They wanted to have all their land together. So they wanted to get rid of the strips. They wanted to enclose the land. Other people were against enclosure. Often three or four people owned many strips in the open fields. If the owners of three-quarters of the land in the village wanted enclosure, they could force it on their neighbours. (In some cases three-quarters of the land might be owned by *one* person.) They went to Parliament in London and asked for an Act of Parliament. They appointed a commissioner. He went to the village. He decided how to divide up the land. When this was done the Act went through Parliament. It was law.

Source B

The Haymakers, 1730. Notice the large numbers of people needed to gather in the hay.

Source A

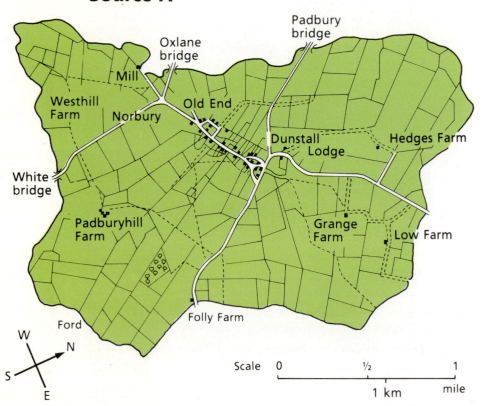

The village of Padbury after enclosure in 1796. The open fields have been divided into individual farms. Each farm has been divided by hedges into smaller fields.

Source C

‘Country people had not been so poor when Sally was a girl. Sally's father had kept a cow, geese, poultry and a donkey cart to carry his produce to the market town. He could do this because he had commoners' rights and could turn his animals out to graze and cut furze for firing.’

Flora Thompson writing about her childhood in the 1880s and 90s. From *Lark Rise to Candleford*, 1947.

Source D

‘More plenty of mutton and beef,
Corn, butter and cheese of the best,
More wealth anywhere (to be brief)
More people, more handsome and prest,
Where find ye – go search any coast,
Than there where enclosure is most.’

From *Five Hundredth Points of Good Husbandry* by Thomas Tusser. This book was popular between 1557 and 1580.

Source E

Case of the petitioners against the Bill for Enclosing Appleby in the early nineteenth century

'There are in Appleby about seventy Freehold Cottages which are entitled to Common right for all manner of cattle in the Open Fields after the Corn is carried and upon the Commons at all times of the Year. Thirty seven of the owners of these cottages and who are most of them also owners of land in the fields think they shall be greatly injured if the Bill passes as it now stands. And therefore desire the following amendments therein:

That two acres of the Commons shall be allotted to and for each and every cottage in Lieu of the right of Common belonging to such cottages.

That the two acres to be allotted shall bear no Share of the Expence of passing the Act but only a proper Share of fencing each two acres only.

That John Willington of Tamworth in the County of Stafford Gentleman be a Commissioner. That the commissioners at this time do make a just Equal reassessment of the poor rate for the said Parish of Appleby and that such assessment of poor rate shall be binding to all persons.

The thirty seven of the owners of cottages applying for the above amendments humbly hope the same will be allowed for these reasons. Because they are not only a majority of the owners but are also a majority of the Freeholders in the parish of Appleby in number though not in value.

And Because the Commons in Appleby consist of about 390 acres of land and the claim of two acres is a moderate claim.'

Source F

On enclosed farms, cows could be kept in separate fields. They could be bred selectively. They could be well fed all the year round because root crops like turnips were beginning to be widely grown on the new farms.

Questions

1 Which two sources see enclosure as damaging small farmers and workers?

2 According to Source E how many acres of land have the 70 freeholders lost if the Commons are enclosed?

3 Look back at Source A on page 4. Compare this with Source A on page 6. Are there more or fewer people living in the village after enclosure? How can you tell?

4 Some historians claim that many small farmers were forced to sell their land and become farm workers. Which sources support this claim?

5 Which sources indicate that enclosure might make people better off? Give reasons for your answer. What sort of people were most likely to become better off?

New Methods

Some people introduced new food crops. Dutch farmers grew turnips for winter animal feed. Some British farmers decided to try this out in Britain. It was much easier to experiment where land had been enclosed. Lord Townshend (1674–1738) had a large farm in Norfolk. He did many things to improve the soil and to grow more food. Other people like Jethro Tull invented new machines. Thomas Coke carried on Townshend's good farming and took special interest in having good farmers as tenants on his land. Between 1778 and 1821 he held sheep-shearing festivals. Up to 7000 people came. In this way new ideas were spread.

Arthur Young was a writer. He wrote many books and articles about farming. He spread the ideas of new methods and discoveries to farmers.

One idea that Townshend used and many, many other farmers copied, was the Four Crop Rotation. This meant that no land had to lie fallow for a year, resting.

Source A

A farm built in the 1820s.

Source B

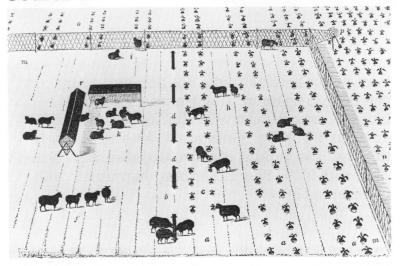

Sheep being fed on turnips out of doors.

Source C

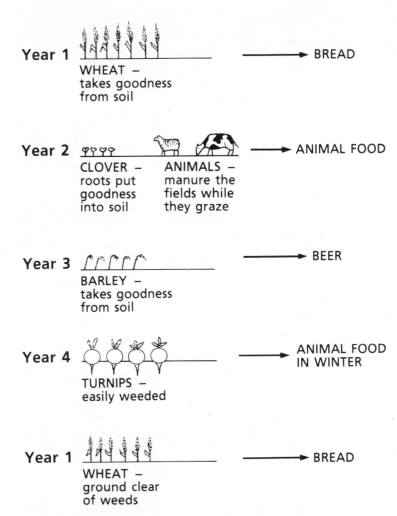

Lord Townshend's Four Crop Rotation. Each year the same field is planted with a different crop. After four years, the rotation starts again with wheat.

Source D

'Great improvements have been made by means of the following:
First by enclosing without the assistance of Parliament.
Second, by use of marl and clay.
Third, by the introduction of rotation of crops.
Fourth by the culture of turnips, hand hoed.
Fifthe, by the culture of clover and grass.
Sixth by landlords granting long leases.
Seventh by the country being divided into large farms. '

From *The Tour of Norfolk* by Arthur Young, about 1770. 'Marl' is a heavy clay. It was used to give 'body' to light sandy soils.

Questions

1 The old farming would have been as follows:

   Year 1   wheat
   Year 2   barley
   Year 3   nothing (fallow)
   Year 4   wheat

   Look at Source C. What were the advantages of the four crop rotation?

2 Look at Source A. Then look back at Source E on page 3. Make a list of the differences between the two farms and their buildings.

3 Why do you think it was much easier for a farmer to experiment where the land was enclosed?

4 What evidence is there in the sources that new methods were being used?

The Improvers

Many people wanted to improve farming. They wanted to grow more food. They invented new machines. They grew new crops. They bred bigger animals. Some people became very famous. Robert Bakewell was born in 1726 in Leicestershire. He travelled around and saw how other people farmed. He took over the family farm when he was thirty. He decided to try out selective breeding. This meant mating the biggest, fattest bulls with the biggest, fattest cows and getting the biggest, fattest calves. In this way the weight of cows doubled between 1700 and 1800. The same thing happened with other animals.

Source A

Cowherd and Milkmaid, painted in the early nineteenth century. Even by 1800 many farm animals were still small, thin and 'leggy'. Improvements spread slowly.

Source B

> 'Bakewell began to let his best rams to other farmers, so that by mating them with their ewes, they could improve the quality of their lambs.'

From *The Story of Agriculture* by J. G. Crowther, 1958.

Source C

Number of people living in England and Wales between 1701 and 1801 (estimated)	
1701	5.8 million
1711	5.9 million
1721	6.0 million
1731	5.9 million
1741	5.9 million
1751	6.1 million
1761	6.5 million
1771	7.0 million
1781	7.5 million
1791	8.2 million
1801	9.1 million

Source D

Many of the new breeds of animals were very big but very fat, so their meat was very fatty to eat. The caption on this painting reads: 'This hog, the property of Mr Charles Butler, was killed at Tidmarsh Farm near Pangbourn, Berks: 15 March 1797'.

Source E

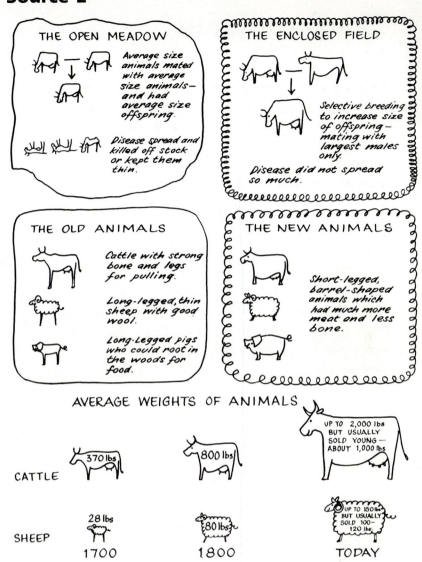

Animal breeders, such as Robert Bakewell, found that enclosing the open fields into smaller, separate fields had many advantages.

Questions

1 Which source represents the old type of farming? Give reasons for your answer.

2 Look at the sources carefully. Then explain how selective breeding worked.

3 Why do you think selective breeding was easier on an enclosed farm?

4 Why do you think other farmers would be interested in and want to try out selective breeding? Look carefully at the sources. Give as many reasons as you can.

The Corn Laws

Britain's population was growing fast. So more corn was needed to feed everyone. Also, in 1793 Britain went to war with France. The war lasted until 1815. France ruled much of Europe. This made it difficult for Britain to import corn from Europe. Several bad summers meant less corn too. So there was a shortage of corn. A shortage of corn meant a shortage of bread. A shortage of bread meant that the price went up. So all through the war the price of bread and corn was high. The high price of corn suited the farmers. Then peace came. More corn came to Britain. The price went down. The farmers protested. They said they were ruined. They insisted Parliament passed a law. This was the new Corn Law of 1815.

Source A

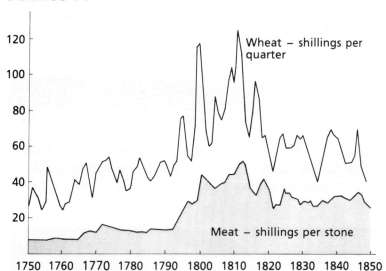

Wheat and meat prices, 1750–1850.
(For Conversion Table turn to page 46.)

Source B

‘A Corn Law (1815) prohibited the import of corn unless the home price rose to eighty shillings (£4) a quarter (about 12.5 kilogrammes).’

From *Britain Since 1700* by R. J. Cootes, 1982.

Source C

Cartoon of about 1820. The artist is commenting on the way in which farmers and dealers kept corn (wheat) back. This made the price of corn go up. Then they would sell it when they thought the price was as high as it would go. The artist is pointing out that the dealer and the farmer are as bad as each other.

Source D

‘There was corn enough for everybody – that was the hard cruel part of it – but those who owned it would not sell it. They kept it back, they locked it up.’

From *Ploughtail to Parliament* by Joseph Arch, 1898.

The Skeleton at the Plough. This cartoon shows the farmworker starving because of his low wages and the high price of bread, made from wheat. The farmer, meanwhile, grows fat.

Source F

‘The immense quantity of foreign corn imported in the year 1814 has so over-stocked the markets that scarcely any sale can be obtained for it in many parts of the country . . . If corn remains at the present low price, the farmer cannot afford to grow it; therefore the labouring poor will starve for want of employment, and the poor rates must be increased.’

From a letter to the Board of Agriculture from Thomas Pilley, a farmer, 1816.

Questions

1 Look at Source B. What did the Corn Law of 1815 say?

2 What arguments does Thomas Pilley put forward for keeping out foreign corn? (Source F).

3 The Corn Law was meant to keep the price of corn high and steady for the farmers. What evidence is there of its success or failure?

4 What does Source C tell you about corn growing and dealing in the early nineteenth century?

5 Which other source or sources support Source C?

6 How could Source C explain Source A?

The Drovers

For hundreds of years the most important long-distance travellers were the drovers. They drove cattle from Scotland and Wales to London. They drove hundreds of cattle at a time. The herds were noisy, dirty and blocked the droving roads for hours at a time. The small black cows from Scotland and Wales were very tough. They walked hundreds of miles. They had about three weeks rest in green fields near London. They got fat. Then they went to market and were slaughtered for meat.

Sheep, pigs, turkeys and geese also went to market along droving roads. It was the only way to get animals to market before railways were built.

Source A

A Scottish droving road. The beginning of the long journey to London.

Source B

Two drovers at the end of the nineteenth century. They went regularly to Ireland to buy cattle and drove them back to Suffolk.

Source C

> 'Halfpenny pastures reflect the cost of grazing per night per beast. The standard charge for a night's accommodation for a drover for the last two hundred years of the droving era, was between 4d (2p) and 6d (2½p) a night.'

From *The Drovers* by Shirley Toulson, 1980.

Source D

> 'Three hundred droves of turkeys pass in one season over Stratford Bridge.'
> (A drove was between 300 and 1000 turkeys. The bridge is on the Ipswich-Colchester Road. The turkeys were coming from Norfolk.)

From *Tour Through England and Wales* by Daniel Defoe, early eighteenth century.

Source E

WHITNEY TOLL BRIDGE ACT, 1797, 36, GEORGE III, CAP, 56, (VI)
TOLLS,

FOR EVERY HORSE, MARE, GELDING, OX, OR OTHER BEAST DRAWING ANY CARRIAGE THE SUM OF FOURPENCE HALFPENNY.   4½d.

FOR EVERY HORSE, MARE, OR GELDING LADEN OR UNLADEN AND NOT DRAWING THE SUM OF TWOPENCE.   2d.

FOR EVERY PERSON OR FOOT PASSENGER THE SUM OF ONE PENNY.   1d.

FOR EVERY SCORE OF OXEN, COWS, OR NEAT CATTLE THE SUM OF TENPENCE.   10d.
AND SO IN PROPORTION FOR ANY GREATER OR LESS NUMBER.

FOR EVERY SCORE OF CALVES, HOGS, SHEEP OR LAMBS THE SUM OF FIVPENCE,   5d.
AND SO IN PROPORTION FOR ANY GREATER OR LESS NUMBER.

A tollbridge in Whitney, Herefordshire. This was where many of the drovers crossed from Wales to England. In Wales, most drovers followed mountain routes, avoiding the roads so they did not have to pay tolls. Whitney was the first place they paid tolls.

Source F

> 'Above all they were responsible for the large sums of money which the cattle represented, on the hoof on the outward journey and homeward in hard cash.'

From *The Drovers* by Shirley Toulson, 1980.

Source G

A drover at the end of the journey, at Smithfield Market, London. His drover's licence is fastened to his arm. From Tudor times all drovers were licensed. They had to be over thirty years of age, householders and married. Having sold the cattle, the drovers walked (or caught a stagecoach, leaving their dogs to walk) home, carrying the money for the cattleowners.

Questions

1 How many different sorts of animals were driven to market according to the sources?

2 What expenses did a drover have on the way south?

3 Study Sources B and G.
  a What differences do you see between the drovers portrayed in the two sources?
  b How do you account for these differences?
  c Which do you think is more true to life? Give reasons for your answers.

4 Why do you think the drovers had to have a licence?

Captain Swing – The Labourers' Revolt

Bread was the main food of working people. So if bread cost a lot, many people were hungry. In 1795 some magistrates in Speenhamland in Berkshire were worried that many poor workers were starving. This was because the price of bread had risen so much. The magistrates decided to pay poor workers some money each week to make their wages high enough to live on. Many other places in the south of England copied this system.

Unfortunately, this meant that many farmers kept on paying low wages. They knew the farm workers would be given extra money. Also the extra money depended on how many children a man had. This encouraged large families. Then there were even more mouths to feed.

By 1830 farmworkers only earned about 7s or 8s (35p or 40p) a week. A loaf of bread cost 1s (5p). They were very, very poor. In 1830 some farm workers in Kent revolted. They demanded another 3d or 6d (1½p or 2½p) a day. They burnt hay ricks and barns. Some rich people were sympathetic. But most were afraid. They crushed the revolt.

Source A

> 'The low levels of wages, the game laws, the degradation of the poor law, the decay of living-in and the growth of rural slums, the immobility of surplus labour and the lack of alternative occupations, together with the loss of winter employment to the threshing machine which led up to the labourers' revolt of 1830.'

From *The Agricultural Revolution 1750–1850* by J. D. Chambers and G. E. Mingay, 1966.

Source B

> 'Came upon three poor fellows digging stone for the roads, who told me that they never had anything to eat but bread and water to wash it down. One of them was a widower with three children; his pay was eighteen pence (7½p) a day; that is to say, about three pounds of bread a day each, for six days of the week: nothing for Sunday, and nothing for lodging, washing, clothing, candlelight or fuel! Just such was the state of things in France at the eve of the Revolution!'

From *Rural Rides* by William Cobbett, 1830.

Source C

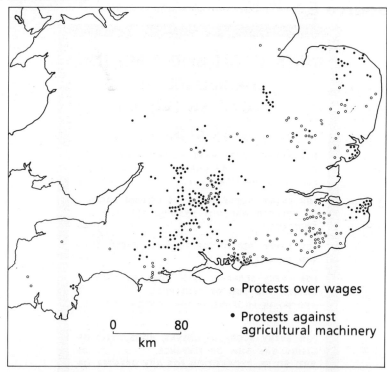

° Protests over wages

• Protests against agricultural machinery

The revolt spread as far west as Dorset and north to Northamptonshire, but the main revolt was in the south east. The farmworkers of the south east of England were particularly poor. The population had risen. There was not enough work on farms. In the north it was different. There was plenty of work in the cotton and wool mills, the iron works and coal mines.

Source D

THE LIFE OF A LABOURER

CONTENT HAVING FOOD & RAIMENT

John Coulter Pauper two shillings weekly

BEGGAR'D BY MISGOVERNMENT AND RECEIVING ALMS OF THE PARISH

Inquire into your distress pho!! nonsense

PETITIONS.

GRIPEALL'S THRASHING MACHINES

IN IGNORANCE TRIES TO RIGHT HIMSELF AND GETS

HANG'D

PUNISHMENT IN ENGLAND FOR A BLOODLESS RIOT.

A series of cartoons about the Labourers' Revolt or Swing Riots of 1830–31. The name Swing Riots comes from the mythical leader of the revolt – Captain Swing.

Source E

'The sentenced men came out looking eagerly at the people until they recognised their own and cried out to them to be of good cheer. "'Tis hanging for me," one would say, "but there'll perhaps be a recommendation for mercy, so don't fret till you know." And another: "Don't you cry, old girl, tis only fourteen years I've got, and maybe I'll live to see you all again." And so one and another as they filed out past their weeping women on their way to the transports in Portsmouth and Plymouth harbours.'

From *A Shepherd's Life* by W. H. Hudson, 1910.

Questions

1 Look at Source A. What reasons does the writer give for the outbreak of the Labourers' Revolt in 1830?

2 Which sources support Source A? Give reasons for your choice.

3 Who does the artist blame for the revolt, in Source D?

4 What punishments did the men receive?

5 Source A mentions no alternative occupations. What alternative occupations were there in the north of England by 1830?

The Labourers' Revolt was crushed. It was 1833. In Tolpuddle in Dorset, a farm worker earned 8s (40p) a week. The farmers said that times were bad. They cut this to 7s (35p). Then they said they could only pay 6s (30p) a week. Farm workers faced starvation. Six men met together. They wanted to form a trade union for farm workers. This trade union could ask for more money. They formed the Tolpuddle Lodge of the Friendly Society of Agricultural Labourers. The local farmers were afraid. But trade unions were legal so they could not stop the union. However, the six men had taken an oath to stand by each other, and to be secret. This was an 'unlawful oath' under an old law. So the six men were arrested and tried in 1834. They were sentenced to seven years transportation to Australia. Many people thought this was terrible. Later the men were pardoned and came back to Britain.

Source A

An earthenware group made in about 1815. Farmers complained frequently that they were too poor to pay their workers any more money. One of the things farmers complained about was the tithe. A tenth of what they earned went to the vicar. This little group shows a well-known joke. The farmer and his wife say: 'you can't have the tenth piglet, the tenth lamb, the tenth bundle of wheat, but you can have the tenth child.' (Many people felt the strain of having too many children to support.) The vicar is not amused.

Source B

A painting of a country cottage.

Source C

'Look at these hovels, made of mud and of straw; bits of glass, or of old cast off windows stuck in the mud wall. Enter and look at the bits of chairs; the boards tacked together to serve for a table; the floor of pebble, broken brick, or of the bare ground; look at the thing called a bed; and survey the rags on the backs of the wretched inhabitants.'

From *Rural Rides* by William Cobbett, 1830.

Source D

'The theft of rabbits, hares, game birds and fish reached astonishing proportions. In 1843 one in four convictions in Suffolk was against the Game Laws. Many of the poachers were young labourers who took a few rabbits but there were also – as in the case of sheep stealing – gangs, often armed and disguised who almost controlled isolated parts of the countryside until the 1860s.'

From *Rural Crime and Protest* by David Jones, 1981.

Source E

'There must always be some of bad character who will not get their livelihood by honest means. These begin by snaring a hare. Like the hound that has tasted blood, this gives them a relish for catching game of which it is impossible to break them.'

From *Gamonia or the Art of Preserving Game* by Lawrence Rawstorne, 1837.

Source F

'The 1830s and 1840s were hard times: poaching and animal stealing increased noticeably once the harvest money had been spent.'

From *Rural Crime and Protest* by David Jones, 1981.

Source G

Working in the fields in winter. A painting by Sir George Clausen entitled *December*.

Questions

1 Look at Sources B and G. What are the two artists trying to show about country life?
2 Does Source C support Source B, or Source G?
3 Looking at all the sources, what sort of people do you think went poaching?
4 From which class of people do you think the writer of Source E came? Give reasons for your answer.

The Repeal of the Corn Laws

Farmers liked the Corn Laws. The Corn Laws stopped corn coming into Britain from other countries. So British farmers felt safer. They got a good price for their corn. But this meant that bread was expensive.

In 1839 the Anti-Corn Law League started. Factory owners and townspeople did not care about the farmers. They wanted cheap bread. So they wanted the Corn Laws repealed.

Richard Cobden and John Bright belonged to the Anti-Corn Law League. They were good speakers. They became Members of Parliament. They pressed the Government to repeal the Corn Laws. Robert Peel was the Prime Minister. In the end he agreed with the Anti-Corn Law League.

Also, in 1845 potato blight destroyed three-quarters of the Irish potato crop. This forced Peel to push through the repeal of the Corn Laws. Then corn could be imported into Ireland.

Source A

Sir Robert Peel, the Prime Minister, who pushed the repeal of the Corn Laws through Parliament. Many people saw him as a cold man. (It was said that his smile was like the silver handle on a coffin.) But he destroyed his own career by repealing the Corn Laws, which was what he felt was right to do.

Source B

'Richard Cobden decided that free trade in corn would solve four problems: First, it would bring prosperity to the manufacturers of all sorts of goods. (If people had to pay less for bread, they could spend more on clothes, carpets and sewing machines.) Second, it would make food cheaper and help the poor. Third, it would make farming more efficient. (If people spent less on bread they could spend more on milk, cheese and meat.) Fourth, foreign countries would sell corn to Britain. Britain would sell goods, such as machines, to them. This would bring peace and prosperity.'

Adapted from *The Age of Improvement* by Asa Briggs, 1960.

Source C

A cartoon of fat farmers in 1849.

Source D

A cartoon by George Cruikshank called *The March of Bricks and Mortar*. It shows the growth of towns at the expense of the countryside.

Source E

Adapted from *The Age of Reform* by L. Woodward, 1962.

Source F

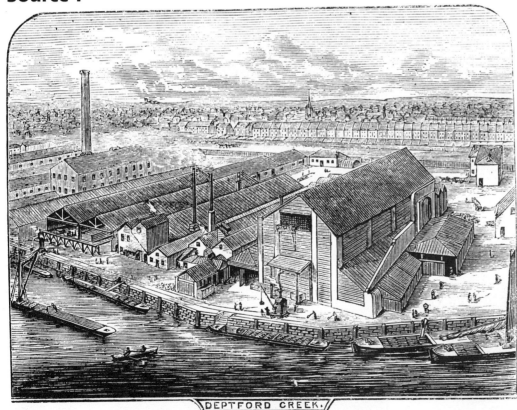

DEPTFORD CREEK.

More and more factories were being built like this one making the new artificial fertilizers for farmers to buy. These fertilizers helped increase the amount of food grown.

Questions

1 What four problems did Richard Cobden say that free trade in corn (repeal of the Corn Laws) would solve?

2 Does Source E agree or disagree with Source B? Give reasons for your answer.

3 Which sources are evidence that more and more people were working in factories and towns?

4 Is the artist of Source C for or against the Corn Laws? Give reasons for your answer.

5 What sort of people, according to Source E, wanted to keep the Corn Laws? Does any other source support this?

6 Read Sources B and E carefully. Why do you think Cobden might say farmers were clodpates?

Women and Children

The men worked hard and long. So did the women. They worked in the home. They had the children. They often worked in the fields too. The women had to make the money stretch to feed the family. The children started to work young too.

Source A

> 'Every morning they were bundled into a piece of old shawl, a slice of food thrust in their hands and they were told to go play while their mothers got on with the housework.'

From *Lark Rise to Candleford* by Flora Thompson, 1947.

Source B

> 'When the corn was in, a mother and her children would go out gleaning. The corn they picked up would be threshed and sent to the miller for grinding. This flour made a family's bread supply secure for a month or two.'

From *The Victorian Countryside* by G. E. Mingay (Ed.), 1981.

Source C

> 'Our food was apples, potatoes, swedes and bread and we drank our tea without milk or sugar. Skim milk could be bought from the farm but it was thought a luxury. Nobody could get enough to eat no matter how they tried. Two of my brothers were out to work. One was eight years old and he got 3s (15p) a week, the other got 7s (35p).'

From *Akenfield* by Ronald Blythe, 1969.

Source D

A young boy crow scaring.

Source E

> 'Many groups of children and women were recruited by a gangmaster to carry out weeding, stone picking, potato setting and other tasks, first on one farm and then on another. Kicking, knocking down and beating were the methods adopted to make the slow and the young children keep up.
>   The Gang Act was passed in 1867. No children under eight were allowed to work in an agricultural gang. All gangmasters must have a licence.'

From *The Victorian Countryside* by G. E. Mingay (Ed.), 1981.

Source F

> 'When the milking was finished they straggled indoors, where Mrs Crick, the dairyman's wife – who was too respectable to go out milking herself – was giving an eye to things.'

From *Tess of the d'Urbervilles* by Thomas Hardy, 1891.

Source H

A woman digging potatoes.

Source G

'My first job was crow scaring and for this I received fourpence (2p) a day. This day was a twelve hour one. There was a smart taste of the farmer's stick when he ran across me outside the field I had been set to watch.'

From the autobiography of Joseph Arch, 1826–1919.

Questions

1 Make a list of the jobs women and children did according to the sources.

2 Does Source D support Source G? Give reasons for your answer.

3 What are the differences between the working conditions in Sources D, E and H? How do you account for the differences?

The Golden Age of Farming

Parliament repealed the Corn Laws in 1846. Yet the price of bread stayed high. This was because there was not a mass of corn waiting to come to Britain. Other countries ate all the corn they grew. British farmers were pleased. They still sold their corn for a good price. Farmers grew richer. They bought new machines. They drained their fields. They started to use new fertilizers.

Source A

Many new reapers were designed in the nineteenth century. Fewer men were needed to cut the wheat than when teams of men cut the crop by using scythes.

Source B

‘In 1838 the Royal Agricultural Society was founded. In 1845 the Royal Agricultural College was founded. In 1840 von Liebig's *Organic Chemistry in its Applications to Agriculture and Physiology* was written. In 1843 the agricultural research station at Rothamsted was founded by Sir John Lawes. Other text books and magazines followed. The 1840s also saw many new machines for such things as drainage. By 1864 an estimated £8 million from the government and public companies had been invested in land improvement.’

From *The Economy 1815–1914* by Trevor May, 1972.

Source C

A new machine for laying drains under the soil. Draining wet land meant better crops could be grown.

Source D

‘While undoing a rick to 'feed' the threshing machine "the wind was coming against us; the prickly dust flew into our faces, our eyes and down our necks if we'd been foolish enough to wear open-neck shirts."’

From *The Worm Forgives the Plough* by J. S. Collis, 1946.

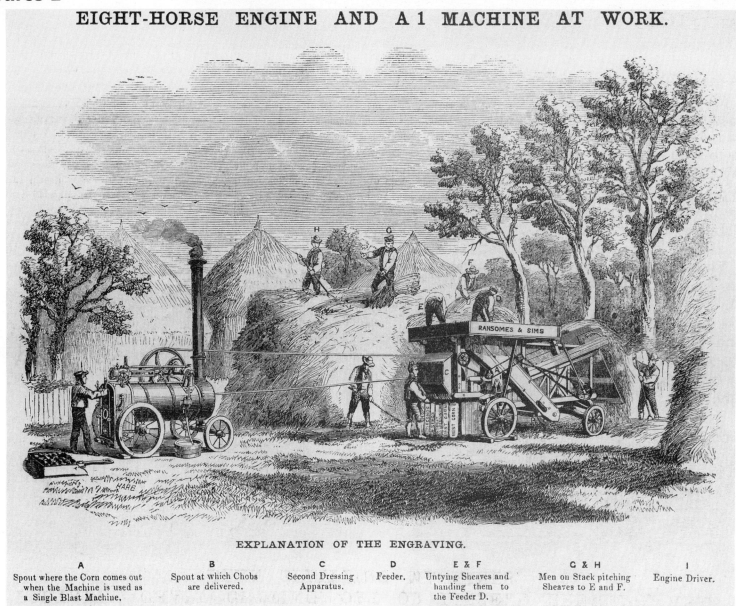

EIGHT-HORSE ENGINE AND A1 MACHINE AT WORK.

EXPLANATION OF THE ENGRAVING.

A	B	C	D	E & F	G & H	I
Spout where the Corn comes out when the Machine is used as a Single Blast Machine.	Spout at which Chobs are delivered.	Second Dressing Apparatus.	Feeder.	Untying Sheaves and handing them to the Feeder D.	Men on Stack pitching Sheaves to E and F.	Engine Driver.

A steam-powered threshing machine of 1860. This separated the seeds of the wheat from the straw. The man who owned the engine would take it from farm to farm when farmers needed their wheat threshed.

Source F

❝Countrymen and women were used to the machinery of windmills and watermills, as well as that of hand or horse driven threshing machines, but for all that it was certainly something when a steam engine first came to work on a farm in a remote village.❞

From *Farming with Steam* by Harold Bennett, 1974.

Questions

1 How did farmers find out about the new research into drainage, fertilizers etc.?

2 What farm jobs are shown or written about in this chapter? Can you work out which other jobs have to be done, even though they are not actually shown?

3 Who do you think put money into drainage and buying new machines, apart from the government?

4 Which sources make it clear that farmers needed factories? Give reasons for your answers.

It was the 1870s. For several years the summers were wet and cold. The wheat harvests were bad. Each year farmers hoped that the summer would be warm. Then they would have plenty of wheat to sell.

They did not notice that something else was happening.

America was a big country. Many people went there to live. They made farms. They grew wheat. The railways carried the wheat to the ports. The steamships carried the wheat to Britain. American wheat was cheaper than British wheat.

Food came to Britain from Australia, New Zealand and South America as well.

Cheap wheat and other crops harmed British wheat growers, but helped farmers who bred animals.

Source B

'In my own country (the Midlands) it was reported in the spring of 1879 that there were over three hundred farms to let amounting to a twelfth part of the entire country. '

From the autobiography of Joseph Arch, 1826–1919.

Source C

An apple orchard in Herefordshire in about 1900. By this time many English people drank wine from France rather than cider.

Source A

A BULL-FIGHT.

Cartoon published in *Punch*, 1881.

Source D

	Number of people moving **from** the country	Number of people moving **to** the towns
1851–61	743,000	620,000
1861–71	683,000	623,000
1871–81	873,000	689,000
1881–91	845,000	238,000
1891–1901	660,000	606,000

From *Home and Foreign Investment 1870–1913* by A. J. Cairncross, 1953.

Source E

> ❝Ten million extra mouths appeared between 1871 and 1901. Real wages rose. Rising wealth usually leads to a rising demand for proteins, for milk and meat.❞

From *The First Industrial Nation* by Peter Mathias, 1969.

Source F

> ❝. . . the sort of man who had bread and cheese for his dinner forty years ago now demands a chop!❞

Agricultural commentator, 1899.

Source G

The Last of England by Ford Madox Brown.

Questions

1 What does Source A suggest was being sold to Britain from foreign countries as well as wheat?

2 How was it preserved from going bad?

3 By how much did the population of Britain grow between 1871 and 1901?

4 Do you think the area mentioned in Source B was a wheat growing area? Give reasons for your answer.

5 Look carefully at Source D. Where do you think the people who 'disappeared' went to? Which source supports this theory?

6 What other imports were coming into Britain?

7 Source E states that 'real wages rose'. Which other source supports this statement?

Farm workers were among the poorest people in Britain. They worked six or seven days a week in all weathers. Townspeople thought farm workers were stupid. They called farm workers 'Bumpkins' or 'Hodges'. Yet these farm workers were people of great skill. They could do many jobs.

In 1872 Joseph Arch founded The Agricultural Labourers' Union. He wanted better pay and conditions for farm workers.

Source A

Joseph Arch in 1872.

'I mounted an old pig-stool, and in the flickering light of the lanterns I saw the earnest up-turned faces of these poor brothers of mine — faces gaunt with hunger and pinched with want — all looking towards me and ready to listen to the words that should fall from my lips. These white slaves of England stood there with the darkness all about them.' Joseph Arch.

Source B

'The day was 7 February 1872. It was a very wet morning. I was busy at home on a carpentering job. My wife came and said: "Joe, here's three men come to see you. What for, I don't know." But I knew fast enough. In walked the three. They said they had come to ask me to hold a meeting at Wellesbourne that evening. They wanted to get the men together and start a Union directly.'

From the autobiography of Joseph Arch 1826–1919.

Source C

My Master and I

'Says the master to me, Is it true what I'm
   told,
Your name on the books of the Unions
   enrolled?
I can never allow that a workman of mine
With wicked disturbers of the peace should
   combine.

I give you fair warning, mind what you're
   about,
I shall put my foot on it and trample it out.
Which side your bread's buttered, I'm sure
   you can see,
So decide now, at once, for the Union or
   me.'

First two verses of a song by Howard Evans in memory of the founding of the Union in 1872, from *The Painful Plough* by Roy Palmer, 1972.

Source D

A shepherd, photographed in Dorset about the turn of the century.

Source E

'He was standing in the ditch leaning heavily upon the long handle of his axe. The continuous outdoor labour, the beating of storms and the hard coarse food had dried him up till his hands were rough and horny and without feeling. His chest was open to the north wind which whistled through the bare branches of the tall elm overhead, blowing his shirt back and exposing the immense breadth of bone and rough skin. forty five he was an old, worn out man.'

From *The Toilers of the Field* by Richard Jeffries, 1892.

Source F

PURSUIT O'. KNOWLEDGE!

First Agricultural (quite a Year after our Branch had been Opened). "WHAT BE THEY POST-ES VUR, MAS'R SAM'L?"

Second Ditto (Wag of the Village). "WHY, TO CARRY THE TELEGRAFT WOIRES, GEARGE!"

First Ditto. "WHAT BE THE WOIRES VUR, THEN?"

Second Ditto. "WHAT BE THE WOIRES FUR? WHY, TO HOOLD UP THE POST-ES, SART'N'Y, GEARGE."!!!

This cartoon appeared in *Punch* magazine on 6 April 1872. Farm labourers were frequently portrayed as stupid and slow by townspeople.

Questions

1 When did Joseph Arch found The Agricultural Labourers' Union?

2 Sources D and E give you some idea of the working days of two men. Why do you think it was difficult to organize men to act together in a trade union?

3 Which source indicates that The Agricultural Labourers' Union would not be very successful? Give reasons for your answer.

The Turn of the Century

Cheap food came from other countries. So farmers had to change. Between 1870 and 1900 they grew only half as much wheat. Luckily more and more people were a little better off. Many of these were town workers. They worked in shops and factories and offices. They had more money to spend. They wanted to buy fresh milk, butter, cheese and vegetables. So many farmers stopped growing corn. They kept cows. They grew vegetables. But even so many farm workers left the farms to work in towns. Many farmers gave up. Many people emigrated. By 1913 Britain grew only one-third of the food British people ate.

Source A

> The cost of bringing one ton of wheat from America to Britain:
>
> in 1860 = 64s (£3.20)
> in 1900 = 16s (80p)

Source B

> ‘At the hotels in which he stayed during his tour of the agricultural districts in 1901–02, Rider Haggard found that the food was mainly of foreign origin, while even the village shops were stocking French and Danish butter, American bacon and tinned meat, Canadian cheese and Dutch eggs and margarine. ’

From *The Economy 1815–1914* by Trevor May, 1972.

Source C

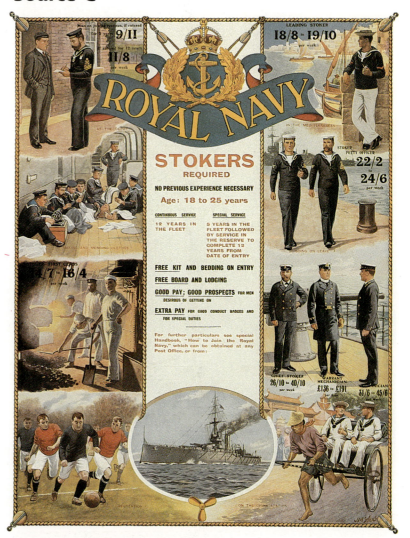

British recruiting poster, 1913.

Source D

> ‘The Richmond Commission (1894–97) listed a number of burdens on the farmer, including tithe and local rates, the new Education Act, increasing labour costs, high rents and unequal railway rates. ’

From *The Economy 1815–1914* by Trevor May, 1972.

Source E

▲ Loading strawberries on to the train bound for London in about 1906.

Source F

◄ Orange box label, about 1898.

Questions

1 From the sources make a list of foreign countries and the foods they were selling to Britain.

2 Explain why Source C would be very attractive to a young farm worker.

3 Apart from cheap wheat from America, what other burdens made a farmer's life difficult according to Source D? Discuss in class how each one might be a burden on a farmer.

In August 1914 Britain went to war against Germany. Most people thought the war would be over by Christmas. But it wasn't. It dragged on and on. More and more soldiers were killed. More and more guns were made. More and more money was spent on fighting the war. Britain and Germany fought on land. They also fought at sea. Both sides wanted to control the sea. Germany decided to try to stop all ships sailing to Britain.

Source A

Source B

Two government posters issued during the First World War.

Source C

> German submarines torpedoed hundreds of ships carrying food to Britain, and at one point the country had only six weeks supply of wheat left. The government wanted more wheat grown. They wanted women to work on the land. With the Corn Production Act of 1917 farmers were guaranteed high prices and stable markets and farm wages were to be kept at a fixed level. Three million more acres of land was ploughed up to grow wheat in two years.

From *All Our Working Lives* by P. Pagnamenta and R. Overy, 1984.

Source D

> Well I came to hiring fair: this side of the street was the place for lads that was wanting hiring. A cattle dealer came up from Bainbridge and looked us over and he picked me out.
>
> "Is ta hiring?"
> "Yes."
> "Work for me?"
> "I will if you give me plenty."
> (I was going to ask ten shilling a week if I was to go away from me own home village.)
> "I'll give thee a pound a week."
> "Aye," I said, "I'll coom."
>
> (I didn't ask what sort of a home I was going to for a pound a week. I could go through Hell for a pound a week!)

From *Country Voices* by Charles Kightly, 1984.

Source E

The Titan 10–20 was introduced in 1914 and in the years 1916–18, nearly 2000 were brought over to Britain. (British factories were making tanks etc.) The Titan was built by the International Harvester Company which was formed by the merger of five leading American machinery manufacturers including McCormick, Deering and Milwaukee Harvester Companies.

Questions

1 Why was it necessary for British farmers to grow more wheat?

2 What did the government do to encourage farmers to grow more wheat?

3 How do you know that America was a leading country for the making of agricultural machinery by 1914?

4 Explain Source B.

5 How do you know there was a shortage of men to work on the farms in the war?

6 Look at Source D. Why do you think farm workers' wages rose during the war?

After the First World War

The war finished in 1918. The British government was giving money to farmers. This meant that farmers carried on growing plenty of food. But all over the world food prices tumbled down. So outside Britain there was lots of cheap food now that the war was over. The government said it could not afford to keep giving money to farmers. So in 1921 the government stopped giving money to help farmers.

Source A

> *Saturday 15 March 1919:*
>
> 'Lyons Hall is now well supplied with farm-hands, so many of the young men who had been called up having come back.'

From *Echoes of the Great War*, the diary of the Reverend Andrew Clark, 1914–19.

Source B

> 'There were 1 million farm workers in 1919. There were 590,000 farm workers in 1930.'

From *All Our Working Lives* by P. Pagnamenta and R. Overy, 1984.

Source C

> 'Farmers were in a very poor way. I remember when we first bought this farm that the value dropped by £1000 in one year. Horses we paid £147 for, one year later they were valued at £47, and we had ten horses. Well, that was a great deal of money. I remember my father coming home. "Look here," he said, "I can't get any money from the bank, I have used my money, I have used your money, I might as well shoot myself."'

From an interview with Cyril Muskett in *All Our Working Lives*, 1984.

Source D

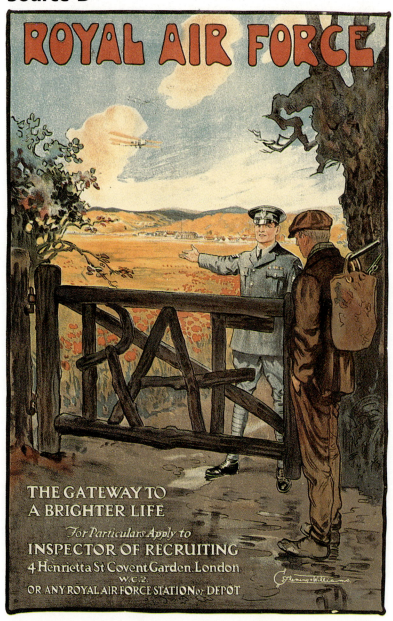

Recruiting poster from the 1920s.

Source E

Although machines such as tractors were being produced in the 1920s, most farmers did not have enough money to buy them. They made do with whatever they could get cheaply. This is an old army lorry with a haysweep on the front.

Source G

> 'The average proportion of the employed population out of work, was roughly six per cent for the thirty years before 1914. It remained in excess of ten per cent from 1922 until 1939.'

From *Britain Yesterday and Today* by W. M. Stern, 1962.

Questions

1 What evidence is there that there were plenty of farm-hands after the war?

2 What evidence is there that farmers were suffering from a lack of money to keep going or buy the new tractors that were available?

3 What did the 40,000 redundant farmworkers (mentioned in Source B) do, according to the other sources?

Source F

BEFORE 1921

cheap food   cheap food

cheap food   cheap food

£ subsidy

AFTER 1921

cheap food   cheap food

cheap food   cheap food

Government money to farmers had kept the price of British food down and kept farmers' jobs safe. Once the government stopped paying money to farmers, British food became very expensive.

Corn prices stayed low. Many farmers were ruined. By 1931 the government realized it must help. All over the world prices were crashing. The government decided to end free trade. It brought in tariffs or duties on foreign wheat. This raised the price of foreign wheat. The government helped farmers in other ways. It gave money to farmers who grew wheat.

This was called a subsidy. It also gave a subsidy to farmers with animals. Then the government set up the Milk Marketing Board. This was successful. So it set up marketing boards for eggs, potatoes, bacon and hops. All these things made farmers more secure. So they started to spend some money on their farms.

Source A

Tithes (a tenth of what a farmer produced) had been changed in 1836 from a payment of an animal or crop, to a rent. In the 1930s farmers were having a bad time and so many campaigned to do away with the ancient payment to the Church.

Source B

Milking bails were developed from the 1920s. They took the milking parlour to the cows, rather than taking the cows to the parlour. The milk was kept in a vacuum from leaving the cow to being deposited in a churn. This meant clean, uncontaminated milk, without the expense of a purpose-built milking parlour.

Source C

'The Milk Marketing Board started in 1933. My father kept bullocks, milked a few cows and mother took butter to Exeter and sold it for about 1s 10d (9p) a pound. We probably took £5 a week. It was not a very good way of making a living. But through the formation of the Milk Marketing Board everybody was paid the same price for milk throughout the country. We were then getting about double the price for liquid milk.'

From an interview with Michael Lee, a farmer, in *All Our Working Lives*, 1984.

Source D

'If a party of farmers are talking shop ask them "does pig farming pay?" Most likely they will say pigs lose £1.00 a piece for fattening. Then someone will say: "more like £1.00 a pig profit." These two opposite opinions are linked by the one word management.'

From *Farmers Weekly*, 1939.

Source E

Fruit and vegetable canning became an important industry in the 1930s. At this factory, the peas were canned, sealed and cooked at the rate of 90 cans a minute.

Questions

1 In what ways did the government help farmers in the 1930s?
2 What evidence is there that farmers helped themselves in the 1930s? How did they do it?
3 What evidence is there in the sources that farmers were hard up in the 1930s?
4 What evidence is there in the sources that farming was a skilled job?
5 Which source points to the way the food industry will go in the future?
6 List any advantages or disadvantages you can think of, to the milking bail system in Source B.
7 Read the first six lines of this chapter. Then copy Source F on page 35, adding a third circle, using the information you have just read in this chapter. Use the heading 'After 1931' for your circle.

The Second World War

The farmers were getting in the harvest when the Second World War started in September 1939. The government had given some help to farmers in the 1930s. But Britain still only produced about one-third of the food everyone ate.

War meant money for soldiers, guns, aeroplanes and ships. Ships took soldiers to fight and ships brought food to Britain. Everyone knew that Germany would use submarines to sink British ships. So Britain had to grow more food. Britain had to grow as much food as possible.

More machines were used. For instance, in the 1930s about one-quarter of all cows were milked by machine. During the war, half the cows were milked by machine.

Source A

> The government told farmers what they were to do. They had to plough up a certain amount of land. Philip Woodward worked on the local War Agricultural Committee:
>
> ❛It wasn't an easy job, because a lot of them didn't want you at all. But you see, they couldn't get any feeding stuff; they couldn't get any petrol; they couldn't get any paraffin, unless it was given to them by the War Agricultural Committee.❜

From an interview with Philip Woodward in *All Our Working Lives*, 1984.

Source B

PLOUGH NOW! *by day and night*

GROW FOOD FOR THE NATION

FEEDING STUFFS FOR YOUR FARMS

KEEP OUR SHIPS AND MONEY FREE

FOR BUYING VITAL ARMS

★THE PRIME MINISTER TO FARMERS AND WORKERS—

The Prime Minister, speaking on February 28th :

"The Minister of Agriculture made a pronouncement last December, when he said: 'If the increase in home production that we want is to be obtained, then the prices must be such as would give a reasonable return to the farmer and enable the farmer to pay a fair wage to the worker.' I want to say again that the War Cabinet endorse that declaration by the Minister of Agriculture."

The government employed many artists to design posters during the Second World War.

Source C

> 'They (the Italian prisoners of war) used to sing when working in the fields.'

From *No Time to Wave Goodbye* by Ben Wicks, 1988.

Source D

> 'The Women's Land Army was started up again just as in the last war. I spent most of the time in forestry.'

Personal memories of Betty Chapman, a land girl.

Source E

> 'It was 16 April 1940. The foreman came out and shook hands. We walked past some acres of fruit trees, 'till we arrived at some ranks of apple trees. Their branches had been cut off and a new kind had been grafted. My job consisted of dragging away and piling up the branches that lay on the ground.'

From *The Worm forgives the Plough* by J. S. Collis, first published 1946. John Collis was 40 years old when he volunteered to work on the land in the war.

Source F

> 'The first week on the farm we all went hop picking, because Mrs Cronish was cooking for ten evacuees instead of her usual hop picking, so we pitched in to help.'

Jean Kircher, an evacuee from London in 1939.

Source G

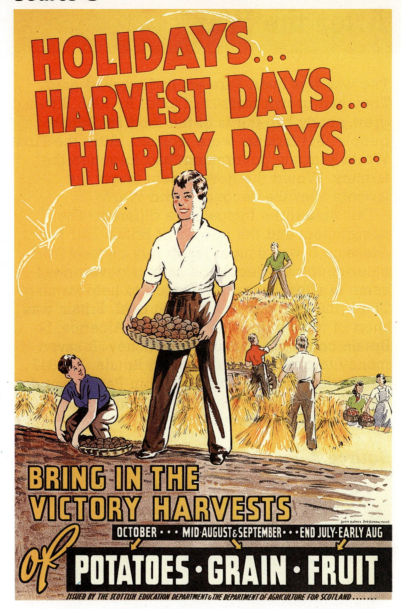

As the war continued there were fewer men to work on the land. Women, children, prisoners of war and holiday makers all helped at harvest time.

Questions

1 Make a list of the different sorts of people who helped on the land in the Second World War.

2 Look at Source B. Why did the government ask farmers to produce more food?

3 What methods did the government use to raise food production in Britain?

The war was over. Farmers had worked hard. They had grown masses of food. By 1945 Britain grew nearly two-thirds of the food for British people. But farmers were worried. After the First World War the government had forgotten the farmers. Would the government forget the farmers after the Second World War?

This time three things were different. The government and the people of Britain had learnt a lesson. It was dangerous to rely on food from other countries. Second, a Labour government came in after the war. It wanted to help farming. Third, the war had cost a great deal. Britain was short of money. It owed money to America. So Britain could not spend much money. It saved money to grow food at home in Britain. In 1947 the Agriculture Act was passed in Parliament. It helped the farmers. The government guaranteed the price of important foods.

Source A

> ❛There were probably 200,000 farms each with their own bull. Now with the Artificial Insemination (AI) that is available you are probably down to 40 or 50 bulls doing most of the work. And the breeding of these bulls is so good that it must improve the sorts of cattle produced. You've got bulls now that cost £80,000, which no ordinary farmer could afford. But you can afford the AI fee of £3 to £5 and for this fee you get the top bulls of the country.❜

An interview with Michael Lee, a farmer, from *All Our Working Lives*, 1984.

Source B

Coming home after a day's ploughing. Horses were still widely used on farms in the 1950s but their numbers dropped rapidly by the 1960s.

Source C

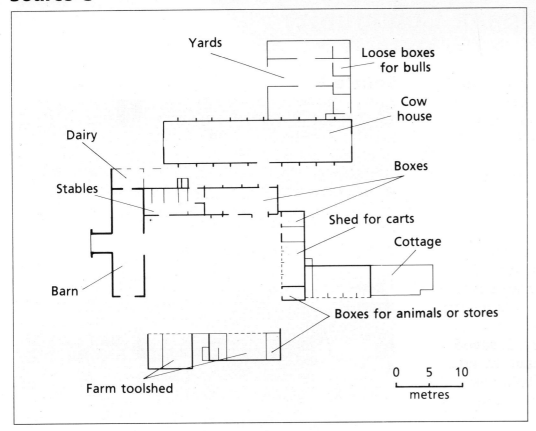

Yards

Loose boxes for bulls

Cow house

Dairy

Stables

Boxes

Shed for carts

Cottage

Barn

Boxes for animals or stores

Farm toolshed

0 5 10

metres

A Dorset farm built in the early-nineteenth century before being modernized in the 1950s.

Questions

1 Why did the farmers get a better deal from the government after the Second World War than after the First World War?

2 Why did farmers stop keeping bulls in the 1950s?

3 List all the changes made in Source D. What is already out of date by the late 1950s, according to Source A?

Source D

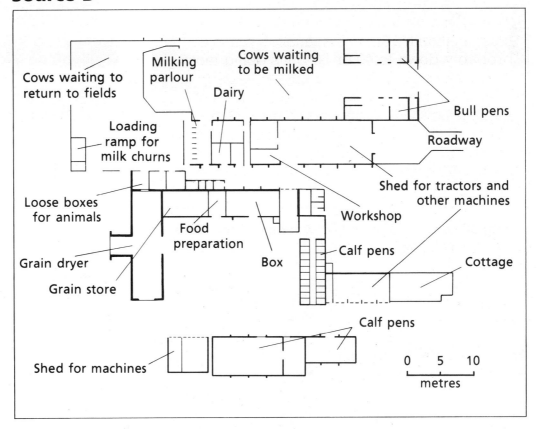

Cows waiting to return to fields

Milking parlour

Cows waiting to be milked

Dairy

Bull pens

Loading ramp for milk churns

Roadway

Shed for tractors and other machines

Loose boxes for animals

Workshop

Food preparation

Box

Calf pens

Grain dryer

Cottage

Grain store

Calf pens

Shed for machines

0 5 10

metres

A Dorset farm after being modernized. Farmers were confident they could sell all the food they grew, so they spent money improving their farms. The government gave them grants to help as well as advice from government experts.

It was 1973. Britain joined the Common Market (which is now known as the European Community or EC). Denmark and Eire joined at the same time. This made nine members altogether.

British farmers had to get used to the Common Agricultural Policy, or CAP. This states that farmers in the EC are guaranteed a minimum price for the food they grow. A tax is put on food coming from *outside* the EC. In Britain some food prices went up. Farmers knew they would get a good price for all the food they produced. The EC bought up extra food stocks to keep farmers in jobs.

Sometimes there were 'mountains' of food. For instance, there was so much butter produced that it had to be stored in huge warehouses. Then it was sold cheaply to countries like Russia to get rid of it.

Source A

Hill farming in Wales. In the valley at the bottom of the picture the farming is quite different.

Source B

'The EC proved less risky than many farmers feared. They had to be prepared to change what they were producing, but with the (EC's) intervention prices could be two or three times higher than prices on the world market.'

From *All Our Working Lives*, 1984.

Source C

'Incomes were bound to suffer as EC member states took action to reduce surpluses of cereals, milk, beef, sheepmeat and other foods. Between 1986 and 1987, 7000 full-time farm workers left the land, mostly because farmers could not afford to keep them.

This year there are signs the surpluses are coming down. In line with the EC, the UK stocks of surplus butter and cereal are reduced and the skimmed milk powder 'mountain' has completely disappeared.

But more farmers will have to give up farming before supplies of food in the EC are back in line with demand. About 1.35 million hectares of land may no longer be needed for agriculture by the mid 1990s, unless there is a desperate world need for food.'

From article entitled 'Farm UK' by John Harvey, published in *Farmers Weekly*, 1989.

Source D

In 1939 there were only about 100 combine harvesters in Britain. By 1973 virtually all grain crops in Britain were cut by combine harvesters.

Source E

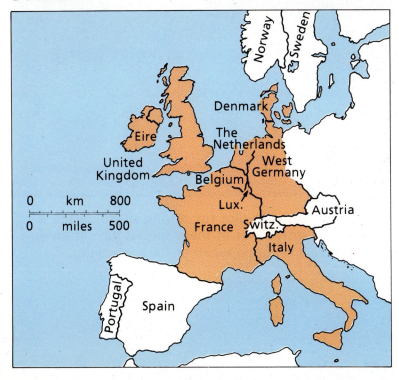

In 1973 there were nine members of the Common Market. Since then, Greece, Spain and Portugal have all joined.

Questions

1 Why did huge stocks of butter or milk build up?

2 Look at Sources A, C and D. Why do you think it might be difficult for a farmer to change what he produces?

3 The EC (Common Market) may not have seemed too risky in 1984. Why was it much more risky for farmers by 1989?

4 Compare the number of men working in Source D with the number working in Source B on page 6.

43

British farmers went on modernizing their farms. They took out hedgerows and made huge fields. They used more and more fertilizers and pesticides. Fewer and fewer farm workers worked on larger and more powerful machines. Animals were raised in buildings that were like factories. The government and the growing British population wanted more food and cheap food.

Then many people began to worry. The EC was producing more than enough food. Modern farming methods got rid of weeds, pests, insects, wild flowers and hedgerows. What else did they get rid of?

Modern methods grew big, lean pigs, cows, sheep, turkeys and geese. Modern methods produced masses of cheap eggs and milk. But what about the animals themselves?

Perhaps we were poisoning the land, wild animals and ourselves in the end.

So slowly people begin to ask for food that is grown without 'poisons' and without cruelty. At the moment we produce plenty of food only because we use pesticides and fertilizers, and all the range of modern methods. If we do not wish to 'kill' the land on which we live, we may all have to produce less food. This has serious implications. How many people can the land in Britain support? How many people can the planet Earth support?

Source A

‘Farmers will drench their land in about 20,000 tonnes of neat pesticides so that we can eat well and make mountains of what we don't need.’

'The Killing Fields', *Radio Times*, May 1989.

Source B

‘Farming is not just farmers and farm workers. You have got all the array of scientists and plant breeders, machinery designers and manufacturers, fertilizer manufacturers and veterinary science.’

From *All Our Working Lives*, 1984.

Source C

‘Taking an average wage, it took just less than 15 minutes to earn enough money to buy a pound of pork in 1988; in 1972 it took more than 26 minutes.’

From *The British Pig*, published by the National Pig Breeders' Association, 1989.

Source D

A modern farm with silos for storing silage. Silage is grass that is stored wet in these special silos. It saves having to rely on good weather for haymaking.

Source E

Building a stack of corn.

e Loaded cart of corn alongside a stack.

f g Sheaves of corn with their butt-ends out-wards.

m Carter forking up a sheaf.

k Field-worker receiving the sheaf with a fork.

h Stacker kneeling on the outside row of sheaves.

i Sheaves of the inside row.

l Sheaf placed most conveniently by the field-worker for the stacker.

Source F

> ❛Most of the ancient hedgebanks have been destroyed by modern farm machinery, but (in Devon) many original hedgebanks still survive, dating back a thousand years and often more.❜

From *One Man's England* by W. G. Hoskins, 1978.

Source G

Keeping hens in confined batteries has meant cheaper eggs in the shops.

Questions

1 Look back at the painting of the farmyard on page 3. Make a list of all the differences between that eighteenth century farmyard and the late twentieth century farm shown in Source D.

2 Why do farmers destroy hedgerows? Discuss in class what uses hedgerows have.

3 'British people are better off today than a few years ago.' Which source or sources support this statement?

4 Discuss in class the ways in which the 'jobs' mentioned in Source B help the farmer.

5 What are the benefits of silos (Source D) as opposed to haystacks (Source E)?

6 Are you prepared to pay more money for food?

Index

Entries in the index refer to written and pictorial sources, as well as the text.

Conversion Table

1 shilling	= 5p
1 quarter	= one-fourth of a hundredweight of wheat, or about 12.5 kilogrammes
1 stone	= 14lbs, or 6.3 kilogrammes